Vine to Vintage:

A Year in the

Life of an Artisan Winery

For my Grandson Eli
Published in 2014 by Hvorfor Ikke, Walla Walla, Washington
USA. © 2013 by Barbara Beito.

Available from Amazon.com, CreateSpace.com
and other retail outlets

Library of Congress Cataloging-in-Publication Data available

ISBN: 978 0 9886155 4 0

Vine to Vintage:
A Year in the
Life of an Artisan Winery

Barbara Beito
© 2013

The Morrison family agreed to let me in on their busy lives for a year, following the vinifera (grapes) from vine to vintage. They make wine that is family grown, family owned and family vinted in the Walla Walla Valley. Wines crafted during the year won gold and silver medals in various competitions.

About the Vineyard:
Morrison Lane Vineyard was begun by Dean and Verdie Morrison with a planting of 4.5 acres of Syrah in 1994 on farmland that has been in the Morrison family for four generations. Over time, other blocks were planted with different varietals for a total of 23.5 acres. The vineyard is located on old stream beds the Cottonwood Creek has wandered over, leaving cobbles and gravel under loess. The federal government designated the Walla Walla Valley as an American Viticultural Area (AVA) and the region was officially recognized by the American Viticultural Association in 1984. Located at Latitude 46° N, Longitude 118.5° W the AVA is a unique growing area that defines the flavors and aromas of locally grown grapes.

About the Winemaker:
Sean Morrison was born and raised in Walla Walla, left town to obtain a degree in Environmental Science with a minor in Chemistry from Western Washington University, and returned to work in the local wine industry for several years. In addition to using grapes from the family vineyard to make wine for Morrison Lane, owned by his parents and specializing in premium wine by varietal, Sean makes wine under custom labels for individuals and other entities.

January and February

The vineyard is prepared for the coming year. Sean begins the year working with wine in barrel from grapes harvested previously.

In the vineyard:
In January the plants are sleeping: spring pruning starts in February.

The focus of dormant pruning is on selecting the best canes and buds. It will impact the fruitfulness of the vine and the ripeness of the grape. Training systems are how the retained canes (and later shoots) shape the vine which impacts how the fruit

1

is exposed to the sun, ease of harvest, etc. Canopy height should be equal to row width.

Dean notes that, other things being equal, early pruning (mid-February) makes the flowers bloom earlier. For example, a grape like Carmenere, among the first to show color and leaf out, will ripen slowly: Dean prunes them early so the fruit will be peaking at harvest time. Other grapes are pruned later (early April) to help them bloom around mid-June.

Morrison Lane Vineyard uses two vineyard training systems depending on which is best for the plant and how the vineyard is managed, to produce optimal yields of quality fruit.

Head trained/cane pruned, for example, is used with the Carmenere, the first vines pruned in the vineyard. A medium diameter cane is selected from near the top of the trunk and used as a horizontal branch called a 'cordon' (canes that are too small or large tend to be less fruitful). The length along the top is kept to four feet (as the plants are set eight feet apart), however many buds that is (in some varietals the buds in the middle are more productive than the buds at either end). All other canes are removed.

If the cordon or trunk has been damaged (due to injury, cold, etc.) you pull a newer cane from below to replace the trunk/cordon. The finger/thumb sized light brown canes with smooth bark seen below grew last year. The trunks of the plants were damaged when a bitter cold came too early last fall, but the fruit that grows on the replacement canes this year will be fine.

After a year or so as the cane grows bigger it becomes the new trunk/cordon.

Cordon trained/spur pruned: this next example is a Syrah with a thick trunk and thick horizontal branches (cordon) that grew in previous years. The canes that grew on the cordon last year are cut off to form 2 bud spurs.

At the winery:
Now is the time for vacations, deep cleaning and getting the tasting room ready to open, as well as 'working the market' (tastings for restaurateurs, and other multiple sale points). Sean is waiting for whites and rosés to mature, tasting, blending, and racking as needed, and getting supplies ready to bottle previous vintages.

The whites and rosé came off the vine in the fall of the previous year. Now the winemaker is ready to put the final rosé blends together and then they are ready to bottle. Rosé is meant to be fresh and youthful – they don't hold up for long so drink them soon!

There are a couple of ways to make rosé. One way is to whole cluster press it, like whites – just press the juice off and pitch the rest. The other way is called the saigneé (pronounced sen-yay) technique, which Sean prefers.

The saigneé technique involves first destemming the grapes (knocking them off their clusters).

The must (loose grapes and juices) is put into bins. Then, the juice for rosé is siphoned off and inoculated with yeast. The brief time with the skins adds color and flavors.

The rest of the must is inoculated with yeast and will go through primary fermentation with the skins on: the yeast will break down the tannins out of the skins and produce the dark color of red wine. (See information about 'crush' in September – October winery observations).

Sean is currently working with three rosés: the Nebbiolo (a varietal) rosé will probably be labeled Morrison Lane. The color is a beautiful salmon-orange, there is only 1 barrel, and it is more acidic than the others. A second rosé is a blend of Rhone varietals: Cinsault, Counoise, Syrah, Viognier and a tiny bit of Grenache. The third rosé contains Italian varietals: Dolcetto, Nebbiolo, Sangiovese and Barbera.

The whites and rosé have just come off 'cold stabilization'. When wine is chilled it sometimes gets cloudy. To be sure it doesn't, the winery chills it down to 35 degrees and the tartrates precipitate out so it won't get cloudy when chilled again in your refrigerator.

There are a couple of ways of using cold to stabilize the wine. Traditionally, you wait until it gets really cold outside, put the barrels out for a couple of weeks and hope it works. You can also put the wine in a jacketed tank, put glycol in the tank liner and chill it down for a couple of weeks.

At this time the red wines are in barrels going through and finishing up secondary fermentation: the malo-lactic culture bacteria (Oenococcus) is converting malate to lactate. As the bacteria works, it breaks carbon atoms off the malic acid, using the energy from that reaction to stay alive and continue the conversion. (There is a rough bench test to check the progress called paper chromatography. Filter paper is rolled into a cylinder and

dots of the sample are put along the bottom. A solvent helps the acids move up the paper through capillary action. As each acid has unique properties, the resulting spots/streaks can be compared to standards.) When the process is complete all that will be left, in addition to flavors and other characteristics, is lactic acid, and secondary fermentation is over. Winery buildings frequently have their own resident Oenococcus colonies but many winemakers like to inoculate to retain control.

About the influence of 'cooperage' (barrel manufacturing): The winemaker uses the barrel as another element in the flavor mix options. The same grape will taste different if it sits in a 3rd year barrel and a 1st year barrel of the same forest, different if it sits in a 1st year toasted barrel or a 1st year barrel that is heavy toasted, and different if it sits in barrels made by the same cooperage but of different wood. The influence of the barrel is over by the 3rd year as flavors have been leached out unless the inside of the barrel is shaved down to new wood.

March – April:

Dormancy ends in the vineyard. Sean continues blending and racking the whites and rosés, getting previous harvest blends put together for bottling this year, and racking and blending the red varietals harvested last year so they get more aging – he will bottle them sometime next year.

In the vineyard:

Dean is pruning vines to select productive buds and canes, planning to prune higher elevations in April, waiting for bud break and hoping to avoid spring drama. In Walla Walla a cold spell after bud break can damage or even end the crop. Once the buds start swelling the dormant period is over and the vine is vulnerable to damage. 'Bud break' is when the swelling is still in a clump but has started to separate into leaves. 26 to 28 degrees is the danger point for bud damage.

Dormant   Bud swell   Bud Break

There are 3 parts to a bud. In general, the primary, on the outside, produces the big clusters of fruit. The secondary, in between, is the back-up system producing clusters but not so large or numerous (as a general rule about 30% of the crop primaries can produce). The tertiary doesn't produce fruit but does produce a shoot. If a freeze kills the primary, the secondary will produce some fruit. If the secondary gets frozen the tertiary produces a shoot. If the tertiary is frozen there will be no shoot or fruit.

Vine drama is different in different places. In Walla Walla fall frosts in September and early October are the major problem as the fruit may not be ripe, or harvested vines have not hardened off yet and the cold can damage them. The Morrison Lane vineyard is along a creek and sometimes the depression pulls cold air along the side of the grapes/vinefera. The year after freeze damage, vines act differently: 'ripe' time changes, etc.

In Walla Walla, spring drama looks like this:

Friday 4/12: Verdie walks me through the vineyard and pictures are taken of different vines in different stages of emergence.

Sunday 4/14: The National Weather Service in Pendleton issues a hard freeze warning for SE Washington

Monday 4/15: A sudden hail storm in the afternoon is followed by a nighttime low of 29 degrees

Tuesday 4/16: The Weather Channel says our nighttime low is 28 degrees

Wednesday 4/17: A visit with Sean who is going to talk with Dean about the vineyard. He had heard it got to 27 degrees for 6 hours last night.

Friday 4/19: the Walla Walla Union Bulletin runs an article by Andy Perdue, Editor of Great Northwest Wine, who noted the spring frost had been cold enough to cause "significant damage" throughout the area. Verdie said a few first leaves looked a little burned (brown at the edge) but the vineyard had not been hurt badly as their vines were just beginning to break.

And another hard freeze came out of nowhere at the end of the month.

At the Winery:
A lot of time is spent working on 'sensory analysis' at first: taste, clarity, color, mouth feel, viscosity, and finish. The winemaker needs to determine the presence (or absence) of desired qualities and figure out production strategies to obtain the characteristics desired. After you get what you want, you do lab/chemical analysis – right before bottling (screening for problems like spoilage organism) and right after bottling. You need to measure alcohol content for the label, pH (you want a 3-4 range), TA (titratable acidity), and residual sugar if you have an 'off dry lot' (if you left the wine a little sweet, like sweet riesling or dessert wine). Technical information is useful at wine events, for wine agents and restaurants, etc.

In the cellar, every couple of weeks there is more racking (d r a i n i n g / cleaning/etc.) and topping off barrels.

'Racking' is anytime you empty a barrel – into another barrel or pump it into a tank, or whatever. Racking takes wine off its solids, and anytime you move wine you leave the solids behind. Racking involves moving the wine out of the barrel, getting all of the lees (dregs, yeast hulls, tartrates and other detritus) out and cleaning the barrels, and then putting the wine back in barrel. You always rack right before bottling.

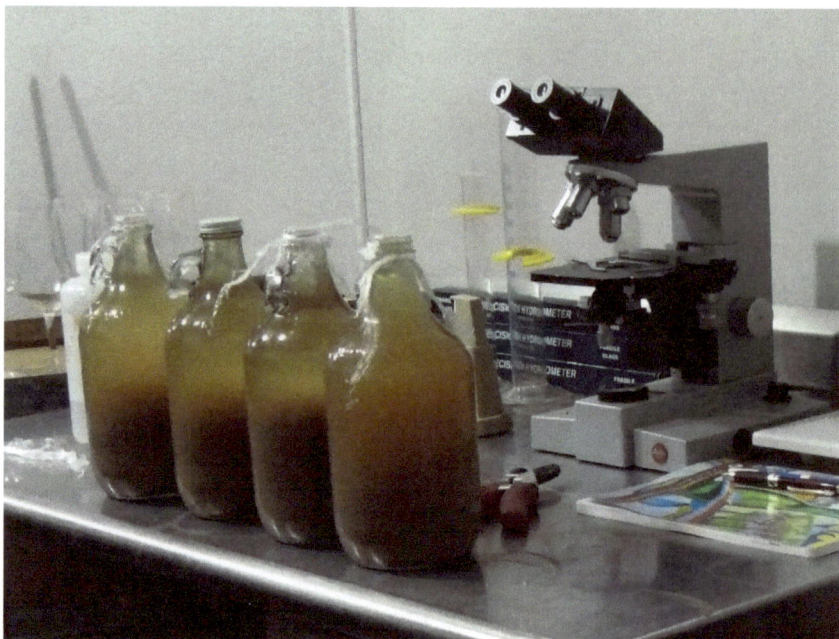

When you get close to the solids at the bottom of the barrel, you can transfer the last of the liquid into a carboy, let the residue settle out over several hours or overnight, and then you will have a little more clear wine to add back into other barrels. This is some Viognier from barrels that have recently been racked. One old school way to check clarity is to light a match and hold it behind a filled clear glass container. How well can you see the flame?

Topping off: anytime you move wine, you taste it – to be sure there is no off taste, etc. (you don't want to spread any problems around) and to help you decide what barrels get topped off with what barrel. Sean's Syrah hard press, picked last November, had very good taste and color so Sean was 'breaking the barrel down' and using it to top off other Syrah.

(Breaking a barrel down is distributing the liquid somewhere else). You don't want to oxygenate wine when topping off – so Sean was careful not to aerate the wine when he transferred it from the barrel into the topping pitcher, and from the pitcher into the barrel. He was using traditional tools. You can also use a topping system with kegs that uses pressurized gas to move wine. The big advantage is that it keeps air away from the wine.

Note: Grapes are tracked according to when and where they were picked. For example Sean has 3 different lots (or pickings) of Viognier plus a 'hard press'. (This means the Viog-

nier by the house was picked in 2 different phases, plus the Viognier at the higher elevation was picked at another time. They are all pressed off separately and put in barrel separately. In addition, after each lot was pressed, the grapes were pressed again harder and the juice from each 'hard press' is collected together and then tracked and barreled separately {as its own lot}.) The hard press naturally isn't as pure as the first pressing. Any/all can be used as part of a blend. When it is picked and crushed is its vintage. Sean uses blue tape on the barrels to track information about the wine inside.

May – June

The vineyard wakes up. Sean determines it is time to bottle some of the wines he has been working on.

In the vineyard:

A week or so into bud break you can find flower clusters, or inflorescence, that look like little tiny grape clusters. These gradually open into tiny wine grape flowers that can be very fragrant. The flowers are wind pollinated. The size of the crop will be influenced by how the flowers develop, are pollinated, and how the fruit sets on.

Inflorescence,   blooming flowers,

grapes getting started.

When the shoots get four to five inches long the spraying starts. Dean and Verdie spray for powdery mildew using Stylet oil to smother the spores. They rotate brands and types of spray so they don't build up resistance in the plants. The 2 most important sprays are systemic sprays just before and just after flowering. Later in the year you spray if you see signs of problems.

In the winery:
The first full weekend in May is Spring Release when wineries offer their most recently bottled wine and 'tasting season' officially starts. This spring, Sean will release a multi-varietal red blend, and pre-release the Rosé of Nebbiolo, the Rhone rosé and a Viognier. Mother's Day weekend brings many Walla Walla families together which means lots of people in town may want to go wine tasting, and beautiful weather draws people out. Even though tasting rooms are busy, production must not stall.

When red wine is through secondary fermentation, you can bottle anytime – it is a sensory evaluation decision. Sean has decided the time is right for some of the wines he is working with. He spends time giving the wine a final filtering before bottling, which is a lengthy process (it isn't unusual for him to spend a night or two at the winery while it's going on).

There are two kinds of filtering systems commonly used (cartridge or plate and frame), and various sizes of spaces in the filters so you can remove bacteria and/or yeast or let some through if you want. Sean takes out both.

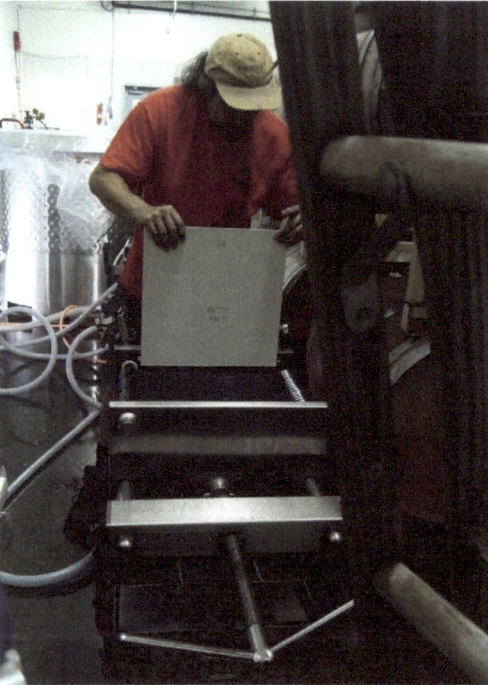

He uses a plate and frame system, which has 19 stacked cellulose filtration pads (cardboard-like material) in frames that are bolted together.

At first water is run through the filters until you can't taste any hint of cardboard or paper. Then it is ok to percolate wine through the system. A tank is hooked up to the filter and wine goes through very slowly, ultimately moving through a diaphragm air pump into a hose and into a different tank where it sits for at least 24 hours. As the pads plug up with particulate, pressure builds so you watch the pressure gauges to see when you need to change anything.

This is an 'after' picture of filtration pads for the Red Wine blend. After the big tanks are settled out it is time for Sean to figure out what order the wines will be bottled in to ensure the process is as smooth as possible.

NOTE: You can't use bleach water to clean equipment because it provides the right conditions for cork spoilage organisms (they like chlorine/bromine/iodine). Equipment is first sanitized with a rinse of sulfur dioxide and then neutralized with a citric acid solution.

Sean has set a mid-May bottling day: everything must be organized and ready because time is money. Sean is ready to process three rosés - the Nebbiolo from 3 different lots for Morrison Lane, the Rhone varietals and the Italian varietals; 3 different lots of Viognier; a red table wine (20% each of Syrah, Cinsault, Counoise, Dolcetto and Viognier); and a Cinsault harvested last fall. Sean wants to bottle when the wine is 58 to 60 degrees Fahrenheit.

The crew feeds bottles into a system that fills, corks, seals and labels them. The process is monitored and

any problems are resolved before boxes are reloaded, labeled and stacked. Sean sets up, provides direction, keeps track of it all and steps in when needed. It's a long, fun day.

Sean's family makes a good team.

Some of today's additions (left) join current offerings in the tasting room. Sean's favorites this year are the Rhone Rosé, the Morrison Lane Viognier, and the Morrison Lane Cinsault.

One interesting decision in bottling has to do with which closure system to use: the 3 main ones are natural cork, synthetic cork or screw tops. Sean likes natural cork or screw tops. One advantage to a natural cork is micro-oxygenation: a very small amount of oxygen comes thru the cork. That is why wine ages: oxygen slowly infuses through the cork and helps the wine reach its peak. Synthetic corks try to mimic this oxygen transfer but Sean is more comfortable with traditional cork. Natural cork also soaks up some of the wine so if the bottle is not left on its side it still seals under most conditions. Synthetics try but the technology is not as proven. Another big consideration: American consumers want to see a natural cork in good wine.

There are different designs to screw caps, and Sean likes them for short lived wines like rosé, Viognier and table wines like the non-vintage red blend when shelf life and aging are not important, as they are meant to be drunk when youthful. The little ring in a screw top is semi-permeable to oxygen and Sean believes it does a better job of mimicking the oxygen transfer process than synthetic cork.

One big win for the winemaker is that due to increased levels of competition among the closure industry, the natural cork manufacturers have developed new processes, and levels of 'cork taint' have significantly decreased, especially over the last ten years. Natural cork contains a naturally occurring fungus that, if it comes in contact with chlorine in the air or a solution, reacts to produce the chemical TCA (tri-chloro-anisole). The term 'corked wine' refers to a wine that is tainted and off-flavor (usually described as tasting of mold) as a result of this reaction.

After bottling the case goods are moved to the chiller room where the temperature can be controlled and consistent. Then it is time to take a few days to relax, clean the production area and grounds, and enjoy a job well done. The new product will sit at least a couple of weeks to recover from 'bottle shock' before it is ready to be poured.

July—August

Watching and analyzing fruit. Selecting barrels of premium varietals and crafting premium blends. The season for tourists, sales and marketing.

In the vineyard:

Dean and Verdie are watching for problems. For example, a wet spell can bring mildew problems, and it is an active time for pests like leaf hoppers or mites. The vines need to be watched for everything as problems can happen all at once. The area's hot days and cool nights are just what the vines like: leaf canopies and berries are growing like crazy. Watering starts around mid-July as Dean wants to stress the vines early to make the berries smaller so they will have a better skin to juice ratio. If there is more skin there will be more color, aroma and flavor.

Grape canopy management is the art and science of figuring out how the leaves can best assist the plant to produce the best possible grapes, given the weather and growing conditions. For the Morrisons, as for others, best is not defined by volume or size of the grapes. In the heat of the summer with temperatures in the low 100s, a canopy that is too open will expose the grapes to possible sunburn. At the same time, the sun needs to hit the clusters to help develop flavors and color, and protect them from powdery mildew. Leaves are removed strategically.

Thinning goes on every 2-3 weeks from July through August (different varietals are on different schedules). 'Shoot thinning' is about the vines, 'green thinning' pertains to clusters. The grapes are most acidic when they are tiny and green: they lose their acidity as they grow. The juice of very young green grapes, called verjus or 'green juice' is very acidic and can be used like lemon juice.

Some vineyards use machines (hedgers) to thin: it chops off the canes a few inches on either side of the support wires. This minimal pruning looks tidy, but encourages branching and the formation of new shoot growth which slows ripening: many times vineyard workers will follow-up with a more targeted thinning. Morrison Lane thins completely by hand, and workers cut or snap off the excess shoots at the cordon so there is no subsequent branching. During the year shoots/foliage are reduced several times to concentrate the grape clusters, thin out the canopy so the sun can penetrate to the grapes, and reduce susceptibility to powdery mildew. Clusters that grow early will diminish shoot growth (or 'devigorate' the vine growth) so you don't want to thin down clusters till later (so you don't get a subsequent bunch of shoots and layers of foliage you will need to thin out).

Cover crops (grown between rows) are important for the beneficial insects that go after the undesirable ones in the vineyard. Currently the Morrisons have a nice diverse cover crop including mustard, clover, legumes, native grasses and the like, and hope for the best.

The vines must be watched for 'veraison' (pronounced vahr-a-soh) or when the grape berries turn color. Along with being beautiful, veraison signals that the plant is moving from berry growth to berry ripening: grapes are softening, taking in sugar and ramping up to harvest. There is about a 2 week window during which the grapes get their color, and although some varietals change sooner than others, they all get their color pretty quickly. The grapes that will be red wine change from bright green to purple, and the grapes that will be white wine change from bright green to yellowish brown (although some varietals will have a little red tint). After the color shift, residual clusters that are green or mixed color are thinned out so all flavors and energy are concentrated into the most ripe fruit.

At the winery:

Sean is working on 2nd and 3rd racking of last year's harvest, selecting his favorite barrels of each varietal, and conducting blending trials (for example, to develop a premium blend of Rhone reds, the very best wine he can make). These blending trials allow experimentation with different ratios and different barrels: the winemaker uses sensory analysis to make decisions. He is starting to put together some blends, and hopes to complete a non-vintage red blend before harvest this fall. This particular Morrison Lane field blend will be reflective of what is in the vineyard: Syrah based and co-fermented with different varietals rather than blended after the fact.

It is a busy time with customers for sales of the wine Sean has made. All of the rosés are selling well in the hot weather, as is the Cinsault, an aromatic light bodied red from big berried fruit.

End of a long day for Sean and winery dogs Wylie and Timber.

Wylie (brown eyes) and Timber (blue eyes) think about a treat

Sean would like to spend more time in the vineyard doing quantitative analysis of how much fruit is out there. One of his future goals is to have the vineyard be a 'certified sustainable' ecosystem through the VINEA/Walla Walla Vinegrowers Sustainable Trust.

Sean is also searching for 'neutral barrels' (3 years old or older) and planning for harvest/crush. How many barrels he will need is based on how many tons of fruit he will have, which requires estimating the yield of various vines. And all this factors in to calculating production costs.

You can estimate scientifically, which involves things like obtaining/sampling cluster counts per acre and cluster weights at different strategic times of the year. You would need to develop a data base specific to fruit at a specific vineyard over years and factor in variables like temperatures, pest problems, etc. You can also guesstimate based on experience, considering the size/density of clusters and how that compares to other years. Estimates vary with the vineyard: varietal(s), planting pattern, training system, soil characteristics and micro-climate are some of the differences that come into play. If you are really good you can estimate to within half a ton per acre. In a sense you are always guessing in the vineyard – how much fruit to drop, how many shoots to pull off and all the rest.

If you underestimate the amount of fruit and consequently don't have enough barrels, you can put the fermenting wine in tanks for up to a couple of months while you round up the barrels you need. Also, you may be able to bottle some finished wine to free up barrels.

<Last ditch effort to get the barrel to seal>

September—October
Fat grapes, thinning and harvest. Crush.

In the vineyard
Preparation for harvest involves vine management related to yield-quality issues. The grape vines are 'stressed' to help concentrate the sugars. For example, water is often withheld, but not entirely, to encourage the vine to focus on ripening the fruit rather than growing shoots and leaves. Sometimes the shoulders are clipped off the grape clusters: Morrison Lane Vineyard does this with some of the grapes grown for other clients. It is a customer preference issue as some believe it helps the cluster concentrate sugars. If there are concerns the sugar levels are not high enough, clusters may be thinned, again to concentrate the sugars. The later in the growing season you do these things the less effective they are going to be.

The summer weather has been perfect for vines. In mid-September the sampled grapes in the vineyard were 20-21 brix (a measure of the sugar in the fruit). Dean says at this time, the fruit usually goes up 1 brix a week: he wants to harvest at 24-25 brix. On and off the following weeks there were a few short but hard rains which had an impact on the fruit. Although there was not a problem with splitting (which tends to happen if rain is steady and prolonged), the absorbed water diluted the fruit bringing the sugar level and the acidity down a bit, postponing the harvest slightly. Time would allow the sugar levels to rise but not the TA (titratable acidity) level. Acidity is measured in the field and at the winery: ideal acidity (Ph around 3.2 – 3.6) acts to stabilize wine and help it age well while resisting spoilage. Adding tartaric acid to increase acidity is pretty standard in hot climates where fast ripening in warm weather can also decrease natural levels in the fruit.

After harvest, grapes need to get well watered to withstand the stress of winter and dormancy.

In the winery
Sean continues racking and blending: in September, preparation for October's crush are underway. This involves maintenance on the press and other equipment, lubing up crush equip-

ment, taking apart and cleaning pumps, giving fittings and hoses an extra cleaning, etc. He has also modified the fermenter tanks so he can rotate them with a fork lift.

Sean is also considering a small bottling run to free up barrels for recycling. That brings up the question 'How do you process a used barrel for recycling/re-use?' First it gets a rinse with really hot water to physically remove as much particulate as possible and open the pores in the wood. Then it is immediately washed with an ozonated water machine to sanitize it. If the barrel doesn't smell right after the Ozone or if it is a problem barrel, you give it a per-oxy-carb soak, then sterilize with a sulfur dioxide solution followed by neutralizing with citric acid.

"Crush" is the term used for the hectic time of year when the fruit of the vines is harvested and pressed for juice. Sean and Dean have been determining in what order varietals are ripening to ensure fruit is picked and arrives at the winery so it can be processed in a timely way. Bins of grapes, weighing about 1,000 pounds each, are hand-picked in the early morning, brought to the winery and set to the side in shade or the chiller room until it is their turn to be processed. The yellow jackets love the aroma and sugar in the grapes!

Sean has decided to whole cluster press fruit for white wine and for rosé. He uses a bladder press.

The press is filled with grapes, closed and rotated a few turns, rolling the fruit around for a quick press. The bladder is then deflated. This is called 'kissing it' and the result is the grapes roll around and pack down a bit so you can put more fruit in the cylinder. The cylinder is topped off and the procedure is started for real: the pressure is brought up on the full cylinder, and juice accumulates in the bottom pan to be pumped out to holding tanks where it is allowed to settle out for a day or so. Sean likes to use a series of slow presses, and while he is pressing the grapes he pulls sample glasses to taste and observe the juice, monitoring color changes. With rosé, Sean is also pressing for color. At some point in the press the juice just starts gushing, as the grapes suddenly give way and give it up.

From this ---   27   to this.

Some batches of fruit are given a 'hard press' when the pressure in the bladder is increased. Again a matter of tasting and analyzing, at some point the taste of the juice changes from sweet and round to sweet with a hard edge. (The sugar stays the same but the pH increases {the acidity drops} producing a buttery mouth feel, and different tannins come out at higher pressures which give an astringent quality.) Sean takes samples of juice during a hard press and tastes to evaluate where things are.

Viognier grapes before

and after pressing!

At this pressing a bin and a half of grapes made about 80 gallons of juice. 240 gallons makes an estimated 100 cases of wine. For estimating field yield: 1 ton of grapes will produce 2-3 barrels, or 50-75 cases (12 bottles to the case). After the fruit is pressed, the grape stems and skins that remain, called pomace (if it's red), are disposed of. The leavings of this Dolcetto, pressed for rosé, may well become compost or food for cattle.

After the juice from the pressed fruit has had a chance to let the particulates settle out into the bottom of the tanks for a couple of days and it gets to a temperature of 55-65 degrees, it is inoculated to start primary fermentation. Different winemakers have their own preferred method of inoculation. Today Sean is inoculating Viognier (a white), a Barbera rosé, and a Rhone blend rosé. Everything needs to be prepped and set up before the process starts because from there it will move quickly, dictated by the temperature of the mix: as it falls closer to the temperature of the juice in the bin a succession of things need to happen.

First samples are drawn from the bins.>

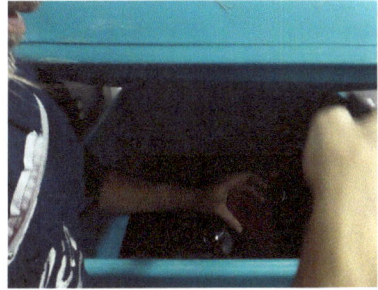

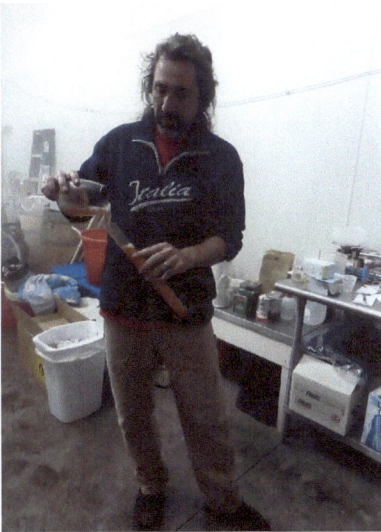

<Samples are tasted for flavor and clarity is checked,

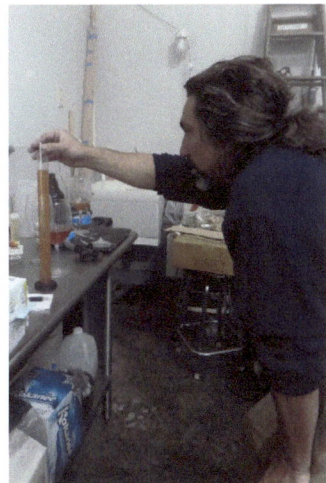

and the brix is measured.>

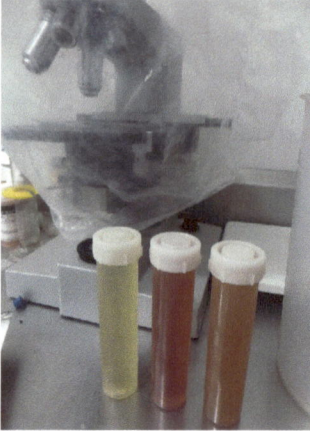

The samples will also undergo chemical analysis (for example pH, malic, and titratable acidity are measured).

<Viognier, Barbera rosé
    and Rhone rosé

Sean then calculates the amount of macro-nutrients (like potassium and nitrogen), water and yeast he will add to each batch. Nutrients contribute flavor to the wine as the yeast converts the compounds into others. He won't add sugar as the brix measures show the juice/must has enough.

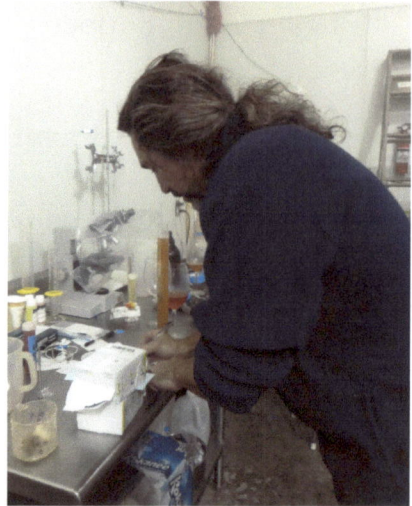

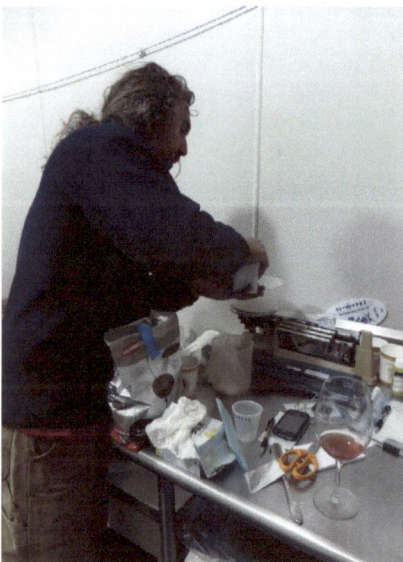

The type of yeast added is different for the different grape varietals because each yeast, as well as each varietal, has its own characteristics. Everything is weighed out. Different winemakers have different recipes.

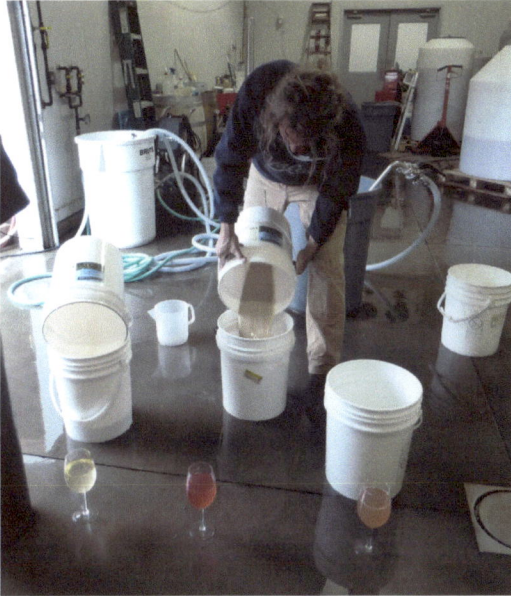

The yeast-nutrients are dissolved in warm water (around 104 degrees), the yeast is added, and then you 'feed it with the must' (a little must is added) to get things rolling.

After set up you continue to feed it every half hour until the temperature of the yeast-mix is within 10 degrees of the starting must in the tank. When the yeast mix is added to the tank is the actual point of inoculation, and the beginning of primary fermentation. Some winemakers don't rehydrate the yeast first, but add everything to the tank and let it go from there.

As the yeast starts converting the glucose and fructose sugars in the must, the temperature starts to build. The yeast is the catalyst: sugar and oxygen in the presence of yeast = ethanol and carbon-dioxide. Temperatures go up and down until the ethanol kills the yeast. Temperatures are checked daily unless one is running too hot or cold: that one is monitored more closely. This primary fermentation lasts from crush until all the fermentable sugar is used up: for whites and rosés typically 3 to 5 weeks. Whites can then age in barrel three to ten months until the winemaker can tell what their taste and character is and decides which barrels to blend.

Red wine is made a little differently. The bins from the field are first emptied into a destemmer which tumbles the grapes and knocks them loose from their cluster. The stems and etc. fall out the end of the machine while the loose grapes are collected in fermentation bins.

The must will go through primary fermentation so the grape skins contribute color and tannin as well as other flavor compounds to the wine. (Tannin also comes from the seeds, as well as the barrel you eventually put the wine in.) At the end of primary fermentation (for reds typically 5 days to 2 weeks) the alcohol content of the wine is almost complete, the flavors have matured to where you want them, and the wine has viscosity ('legs') and tannin.

After primary fermentation is completed, the wine needs to come off the skins before flavors become harsh, astringent etc. The tanks are 'drained down': the 'free run' is pumped out and the must is removed, discarding the seedy stuff in the bottom.

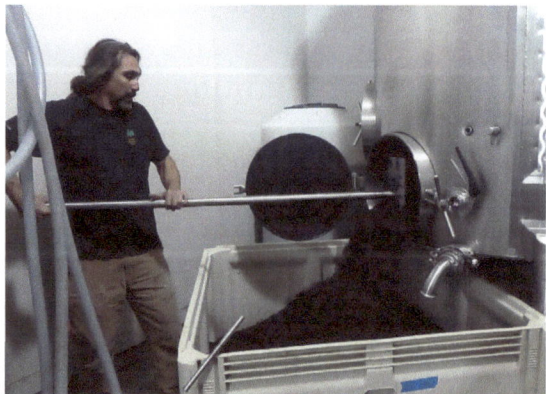

The must is then loaded into the press and the resulting 'press off' juice is added back to the free run. A hard press is also done which will be tracked and managed separately.

Whether white, rosé or red, after primary fermentation is completed the wines are racked into barrels and/or tanks, which of course will require sampling.

November—December

The vineyard slides into dormancy. Secondary fermentation management of reds and flavor development of the wines in barrel. End of the year chores like inventory, accounting, etc.

In the vineyard:
Late fall brings short days and chilly nights. If lucky, the vines will ease into winter. If a freeze comes before harvest is through leaf cells can rupture which means the leaves fall and the fruit can't ripen anymore: the 'raisining' grapes may later be harvested for ice wine. A hard freeze can damage the stems, making the grape clusters fall off the vines.

Dean is watching the vines near the creek side of the vineyard. In Walla Walla and elsewhere cold air stays in lower spots which can be a problem for vineyards planted in lower elevations. Dean has built a berm between the creek and the vines to channel cold air away from the plants, and installed a three story wind machine to pull warm air from 30-60 feet above to mix with the cold air on the ground when there is an inversion and cold air is trapped. As the season changes some varietals, like these Carmenere, really put on a show of color.

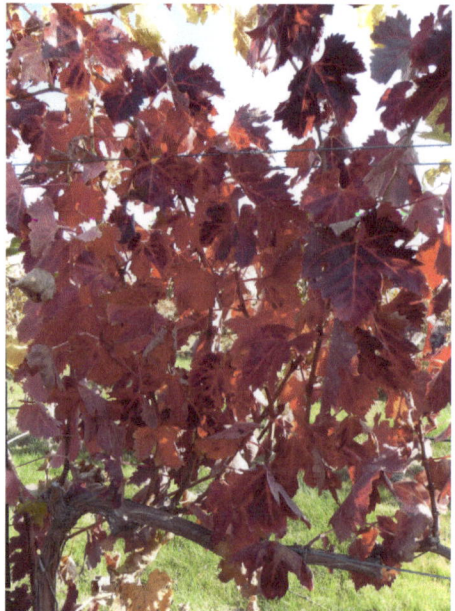

Canes are left on the vines through the winter. The vines go dormant once the leaves are gone – it is a process, the plants 'harden off' and as they do they're more able to withstand cold temperatures for a short duration (maybe five below for a short time). Ten degrees below for an extended time will wipe out the primary layer of the dormant bud and all. The worst case is an early hard freeze like a few years ago: zero degrees in November. The phloem inside the trunk freezes and splits. The plant is most likely dead but sometimes it can come back from below the ground.

Currently the Morrison's Vineyard grows:

Varietal	Rhone	Bordeaux	Italian
Syrah	x		
Counoise	x		
Cinsault	x		
Viognier	x		
Grenache	x		
Carmenere		x	
Petit Verdot		x	
Malbec		x	
Nebbiolo			x
Sangiovese			x
Barbera			x
Dolcetto			x

This year in mid December there is a cold snap and temperatures stay in the single digits for over a week. The wind machine is turned on to pull the cold air out of lower pockets and the Washington State University Extension website is consulted to see damage expectation for different varietals at the current stage of dormancy. It isn't too bad – the Italian varietals can be expected to lose ten percent at temperatures of two to three degrees at this time of year. As Dean says, it will mean a little less thinning in the spring.

November and December is also the time for making Ice Wine. Just like it sounds, the grapes are left on the vine until they are frozen and not too dehydrated. They are harvested when the temperature is fifteen to twenty-five degrees or so. When they are pressed (cluster press like whites) the juice is more concentrated because water remains as ice crystals in the grape and is discarded: less juice, more sugar. You can also press stored frozen fruit. When the yeasts use up all the sugar they can and then die off, there is still sugar left so the wine is high alcohol (about 14%) and sweet.

Late harvest wines are sweet because the acidity is lower and the grapes contain more sugar, usually starting out at over 30 brix rather than 25-26. They are typically not allowed to freeze on the vine.

During the dormant season, some growers dissect buds, put them under a microscope and count 'cluster primordia' of different vines to start estimating crop yield potential. This is possible because by the time vines enter dormancy, the main branches of flower clusters are formed inside the buds that are visible on the canes that grew that year.

In the winery:

The first full weekend of November is Fall Release in Walla Walla and wineries are busy offering a sample of newly released wines regardless of when they were bottled. For example, reds {especially the Syrah and Dolcetto} need to sit a few months in the bottle to recover from 'bottle shock' before being made available. So wines may be in their bottles, but not ready to be released to the public.

It is a busy weekend for the many wineries in the Walla Walla Valley.

In December the busy time for tasting rooms closes out after Barrel Tasting weekend. Barrel Tasting weekend is traditionally an opportunity for wine-makers to show what they have in the barrels – these are not finished wines, and in Sean's case blending has not taken place as he usually likes to ferment varietals separately, choosing some to re-main by themselves and others to blend.

Sean samples the barrels with brother Dan, the original winemaker for Morrison Lane, 2002-2005.

At this time of year Sean's focus is on fermen-tation. In the first part of November Sean inoculates his last bins of fruit: Barbera, Counoise, Carmenere, Cabernet, Rosé of Nebbiolo and a Rhone rosé (the Nebbiolo and Rhone rosé were both made using the saigneé technique described on page 4 and 5). The winery maintains a classic aroma as the yeast con-verts the glucose and fructose sugars in the must into alcohol. When the tops of the bins are lifted you can see the bubbles produced by fermentation and hear the crackle of the yeast as it works.

For a ballpark measure to see if primary fermentation is finished in various barrels, you look for negative brix on the hydrometer. As these bins of wine finish primary fermentation, others are further along in the cycle.

NOTE: "No Sulfites added" does not mean no sulfites are in the wine, as grapes naturally contain a certain amount of Sulfite. Frequently more is added to protect the wine against oxidation and spoilage as it inhibits bacteria and combines with other things in the wine (making 'bound' Sulfite).

The next step for reds starts after 'press off' when the wine is in a tank or barrel. The red wine is inoculated with Oenococcus for secondary fermentation, which will address more subtle aspects of the wine like acid balance and mouth-feel (the buttery texture). In this process bacteria converts malic acid to lactic acid, and can take from two weeks to two or three months, typically finishing up in February (see January-February discussion, Page 6). The winemaker manages fermentation, mostly through temperature control and supplementing nutrition, to encourage the yeast and to uncover or bring forward flavors and other potentials of the grapes.

Temperatures are watched carefully during fermentation, and winemakers have their own targets. Sean likes to keep fermentation 70-80 degrees for reds and charts temperatures to keep track of what is going on. If temps climb over 90 degrees, flavors change: for instance a fresh cherry can become a cherry jam. But there are different tastes for different people to enjoy, and different winemakers encourage different flavor and characteristic development. To cool the fermentation down you can set the fermentation bin outside at night to cool, or you can use chilled jacketed tanks.

After secondary fermentation, reds enter the aging process in barrel, which takes from 6 months to 4 years – you want to hold off on blending decisions as long as you can to allow the wine to develop. Finished wine will be tested again, for residual sugar, alcohol content, etc. Dessert wines and other sweet wines have more residual sugar, usually fructose. (Wine merchants want the tech sheets).

The Wine Club membership list is growing, and parcels are put together for pick up or mail out. Of course there is always inventory to do and competitions to enter.

In both the winery and the vineyard:
Winemakers like Sean and growers like Dean are taking advantage of the pending lull to plan and prepare for the next year.

Next year's crop at rest

Tasting Notes

Sample Tasting Notes:

Winery /Tasting Room name and address:
Morrison Lane Winery   Tasting Room/Library
1249 Lyday Lane   4th & Main Sts
Walla Walla, WA 99362

Brand/label name:
Morrison Lane 2013 Viognier. This wine won Gold Medals at the 2014 International Craft Wine Awards Competition, and at the 2014 Walla Walla Valley Wine Competition.

Vintage   *2013*   Percent alcohol   *14.5*

Varietal or varietals in blend   *100% Viognier*

Where did the grapes come from? How old are the vines?
The fruit is from Morrison Lane Vineyard in Walla Walla, from vines planted in 1996 and 2000.

Who is the winemaker?   *Sean Morrison*

Sensory analysis: (Color/Appearance, Nose, Palate {describe mouth-feel}, flavor descriptors, finish (and faults if any)
The wine is pale wheat-straw in color, clear and has good viscosity. I can smell apricots and pineapple, a hint of citrus and a little bit of spice. A sip feels creamy in my mouth. I can taste light notes of tropical (mango, pineapple) and stone fruits (apricot, peach) but it tastes crisp, not fruity or sweet—it's a very dry wine. There is a hint of clove also, and mineral (like Perrier). After a swallow I am left with a nice balance of acidity, fruit and spice.

Do you love it?   *Absolutely!*

Tasting Notes:

Winery /Tasting Room name and address:

Brand/label name:

Vintage: Percent alcohol:

Varietal or varietals in blend

Where did the grapes come from? How old are the vines?

Who is the winemaker?

Sensory analysis: (Color/Appearance, Nose, Palate {describe mouth-feel}, flavor descriptors, finish and-faults if any)

Do you love it?

Tasting Notes:

Winery /Tasting Room name and address:

Brand/label name:

Vintage: Percent alcohol:

Varietal or varietals in blend

Where did the grapes come from? How old are the vines?

Who is the winemaker?

Sensory analysis: (Color/Appearance, Nose, Palate {describe mouth-feel}, flavor descriptors, finish and-faults if any)

Do you love it?

Tasting Notes:

Winery /Tasting Room name and address:

Brand/label name:

Vintage: Percent alcohol:

Varietal or varietals in blend

Where did the grapes come from? How old are the vines?

Who is the winemaker?

Sensory analysis: (Color/Appearance, Nose, Palate {describe mouth-feel}, flavor descriptors, finish andfaults if any)

Do you love it?

Tasting Notes:

Winery /Tasting Room name and address:

Brand/label name:

Vintage: Percent alcohol:

Varietal or varietals in blend

Where did the grapes come from? How old are the
vines?

Who is the winemaker?

Sensory analysis: (Color/Appearance, Nose, Palate
{describe mouth-feel}, flavor descriptors, finish and
faults if any)

Do you love it?

www.ingramcontent.com/pod-product-compliance
Lightning Source LLC
Chambersburg PA
CBHW041218270326
41931CB00001B/28

* 9 7 8 0 9 8 8 6 1 5 5 4 0 *